Animal Tracks
of Ontario

ANIMAL TRACKS

OF ONTARIO

IAN SHELDON

THE PUBLISHER: LONE PINE PUBLISHING

1808–B Street NW, Suite 140	10145–81 Avenue
Auburn, WA 98001	Edmonton, AB T6E 1W9
USA	Canada

Lone Pine Publishing Website: http://www.lonepinepublishing.com

Canadian Cataloguing in Publication Data

Sheldon, Ian, (date)
    Animal tracks of Ontario

    Includes bibliographical references and index.
    ISBN 13: 978-1-55105-109-3
    ISBN 10: 1-55105-109-5

    1. Animal tracks—Ontario—Identification. 2. Animal tracks—Great Lakes
Region—Identification. I. Title.
QL768.S525 1997   591.47'9   C97-910903-5

Senior Editor: Nancy Foulds
Editor: Volker Bodegom
Production Manager: David Dodge
Design, layout and production: Volker Bodegom, Gregory Brown
Technical review: Donald L. Pattie
Animal illustrations: Gary Ross, Horst Krause, Kindrie Grove
Track illustrations: Ian Sheldon
Cover illustration: Gary Ross
Map: Volker Bodegom
Scanning: Elite Lithographers Ltd., Edmonton, Alberta, Canada

The publisher gratefully acknowledges the support of Alberta Community
Development and the Department of Canadian Heritage.

PC: P6

CONTENTS

INTRODUCTION

If you have ever spent time with an experienced tracker, or perhaps a veteran hunter, then you know just how much there is to learn about the subject of tracking and just how exciting the challenge of tracking animals can be. Maybe you think that tracking is no fun, because all you get to see is the animal's prints. What about the animal itself—isn't that much more exciting? Well, for most of us who don't spend a great deal of time in the beautiful wilderness of Ontario, the chances of seeing the majestic Moose or the fun-loving River Otter are slim. The closest we may ever get to some animals will be through their tracks, and these can inspire a very intimate experience. Remember, you are following in the footsteps of the unseen—animals that are in pursuit of prey, or perhaps being pursued as prey.

This book offers an introduction to the complex world of tracking animals. Sometimes tracking is easy. At other times it is an incredible challenge that leaves you wondering just what animal left those unusual tracks. Take this book into the field with you, and it can provide some help with the first steps to identification. Prints and tracks are this book's focus; you will learn to recognize subtle differences for both. There are, of course, many additional signs to consider, such as scat and food caches, all of which help you to understand the animal that you are tracking.

Remember, it takes many years to become an expert tracker. Tracking is one of those skills that grows with you as you acquire new knowledge in new situations. Most importantly, you will have an intimate experience with nature. You will learn the secrets of the seldom seen. The more you discover, the more you will want to know, and by developing a good understanding of tracking, you will gain an excellent appreciation of the intricacies and delights of our marvellous natural world.

How to Use this Book

Most importantly, take this book into the field with you! Relying on your memory is not an adequate way to identify tracks. Track identification has to be done in the field, or with detailed sketches and notes that you can take home. Much of the process of identification is circumstantial, so you will have much more success when standing beside the track.

This book is laid out so as to be easy to use. There is a quick reference appendix to the tracks of all the animals illustrated in the book on p. 136. This appendix is a fast way to familiarize yourself with certain tracks and the content of the book, and it guides you to the more informative descriptions of each animal and track.

Each animal's description is illustrated with the appropriate footprints and the styles of track that it usually leaves. While these illustrations are not exhaustive, they do show the tracks or groups of prints that you will most likely see. Where there are differences in orientation, left prints are illustrated. You will find a list of dimensions for the tracks, giving the general range, but there will always be extremes, just as there are with people who have unusually small or large feet. Under the category 'Size' (of animal), the 'greater-than' sign (>) is used when the size difference between the sexes is pronounced.

If you think that you may have identified a track, check the 'Similar Species' section for that animal. This section is designed to help you confirm your conclusions by pointing out other animals that leave similar tracks and showing you ways to distinguish among them.

As you read this book, you will notice an abundance of words such as 'often,' 'mostly' and 'usually.' Unfortunately, tracking will never be an exact science; we cannot expect animals to conform to our expectations, so be prepared for the unpredictable.

Tips on Tracking

As you flip through this guide, you will notice clear, well-formed prints. Do not be deceived! It is a rare track that will ever show so clearly. For a good, clear print, the perfect conditions are slightly wet, shallow snow that isn't melting, or slightly soft mud that isn't actually wet. Needless to say, these conditions can be rare—most often you will be dealing with incomplete or faint prints, where you cannot even really be sure of the number of toes.

Should you find yourself looking at a clear print, then the job of identification is much easier. There are a number of key features to look for: Measure the length and width of the print, count the number of toes, check for claw marks and note how far away they are from the body of the print, and look for a heel. Keep in mind other, more subtle features, such as the spacing between the toes and whether

they are parallel or not, and whether fur on the sole
of the foot has made the print less clear.

When you are faced with the challenge
of identifying an unclear print—or even if
you think that you have made a successful
identification from one print alone—look
beyond the single footprint and search
out others. Do not rely on the

dimensions of one print alone, but
collect measurements from several prints to get
an average impression. Even the prints within one track can
show a lot of variation.

Try to determine which is the fore print and which is the
hind, and remember that many animals are built very differently from humans, having larger forefeet than hind feet.
Sometimes the prints will overlap, or they can be directly
on top of one another in a direct register. For some animals,
the fore and hind prints are pretty much the same.

Check out the pattern that the prints make together
in the track, and follow the trail for as many paces as is

necessary for you to become familiar with the pattern. Patterns are very important, and can be the distinguishing feature between different animals with otherwise similar tracks.

Follow the trail for some distance, because it may give you some vital clues. For example, the trail may lead you to a tree, indicating that the animal is a climber, or it may lead down into a burrow. This part of tracking can be the most rewarding, since you are following the life of the animal as it hunts, runs, walks, jumps, feeds or tries to escape a predator.

Take into consideration the habitat. Sometimes very similar species can be distinguished only by their locations—one might be found on the riverbank, while another might be encountered only in the dense forest.

Think about your geographical location, too. Some animals have a limited range, perhaps only in remote parts of the northern part of Ontario. This consideration can rule out some species and help you with your identification.

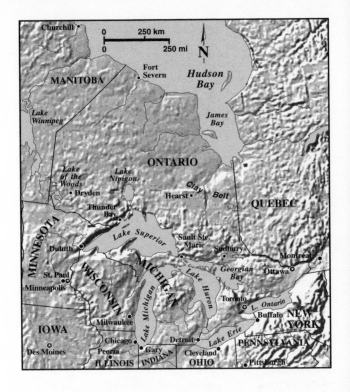

Remember that every animal will at some point leave a print or track that looks just like the print or track of a completely different animal!

Lastly, keep in mind that if you track quietly, you might catch up with the maker of the prints.

13

Terms and Measurements

Some of the terms used in tracking can be rather confusing, and they often depend on personal interpretation. For example, what comes to your mind if you see the word 'hopping'? Perhaps you see a person hopping about on one leg, or perhaps you see a rabbit hopping through the countryside. Clearly, one person's perception of motion can be very different from another's. Some useful terms are explained below, to clarify what is meant in this book, and where appropriate, how the measurements given fit in with each term.

The following terms are sometimes used loosely and interchangeably—for example, a rabbit might be described as 'a hopper' and a squirrel as 'a bounder,' yet both leave the same pattern of prints in the same sequence.

Bounding: Can be used interchangeably with 'hopping' or 'jumping'; often used for patterns that cover short distances; hind prints usually registering ahead of fore prints.

Galloping: Used for the motion made by animals with four even legs, such as dogs, moving at speed; hind prints registering ahead of fore prints.

Hopping: Similar to bounding; usually indicated by tight clusters of prints; fore prints set between and behind the hind prints.

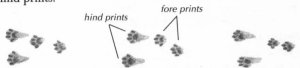

Running: Like galloping, but often applied generally to animals moving at speed.

Trotting: Faster than walking, slower than running.

Other tracking terms:

Alternating track: A left-right sequence, as made by humans walking; often a double register for four-legged animals, which are described as 'diagonal walkers.'

double register   direct register

Dewclaws: Two small, toe-like structures set above and behind the main foot of most hoofed animals.

dew claws

dragline

Direct register: The hind print falls directly on the fore print.

Double register: The hind print overlaps the fore print only slightly or falls beside it, so that both can be seen at least in part.

Draglines: The lines left in snow or mud by a foot or the tail dragging over the surface.

Gallop group: A cluster of four prints made at a gallop, usually with hind prints registering ahead of fore prints.

Height: Taken at the animal's shoulder.

Length: The animal's body length from head to rump, not including the tail, unless otherwise indicated.

Lope: A collection of four prints made at a fast pace, usually falling roughly in a line.

Print: Fore and hind prints are treated individually; print dimensions are length (including claws–maximum values may represent occasional heel register for some animals) and width; together the prints make up a track.

Register: To leave a mark–said about a foot, claw or other part of an animal's body.

Retractable: Describes claws that can be pulled in to keep them sharp, as with the cat family; these claws do not register in the prints.

Sitzmark: The mark left on the ground by animal falling or jumping from a tree.

Straddle: The total width of the track, all prints considered.

Stride: For consistency among different animals, the stride is taken as the distance from the centre of one print (or print group) to the centre of the next one. Other books may use the term 'pace.'

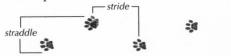

Track: A pattern left by a series of prints.

Trail: Often used to describe a track at length; think of it as the path of the animal.

MAMMALS

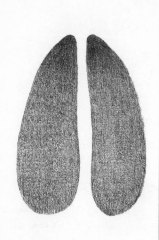

Fore and Hind Prints
Length: 2.0–3.5 in (5.1–8.9 cm)
Width: 1.6–2.5 in (4.1–6.4 cm)

Straddle
5.0–10 in (13–25 cm)

Stride
Walking: 10–20 in (25–51 cm)
Jumping: 6.0–15 ft (1.8–4.6 m)

Size (buck>doe)
Height: 3.0–3.5 ft (91–110 cm)
Length: to 6.3 ft (1.9 m)

Weight
120–350 lb (54–160 kg)

walking   *gallop group*

WHITE-TAILED DEER
Odocoileus virginianus

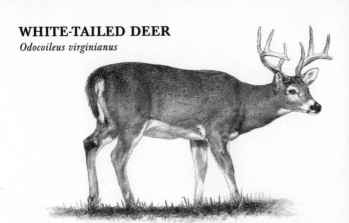

The keen eyesight of this deer guarantees that it knows about you before you know about it. Frequently, all we can see is its conspicuous white tail as it gallops away, which earns this deer its other name of 'flagtail.' These adaptable deer may be found in small groups at the edges of woodlands, often venturing out into open areas. They are widespread through all but extreme northern Ontario, and are frequently seen in farmland and close to residential areas.

Deer prints are heart-shaped and pointed, in an alternating track with hind prints direct- or double-registering on the fore prints. In snow, or when a deer gallops on soft surfaces, the dewclaws register. This flighty deer gallops in the usual style, with the hind prints falling in front of the fore prints; the toes spread wide for better footing.

Similar Species: Young Moose (p. 24) prints may be similar in size; adults leave much longer, wider prints. Elk (p. 22) leave similar, larger prints, but have a limited range.

Fore and Hind Prints
Length: 3.2–5.0 in (8.1–13 cm)
Width: 2.5–4.5 in (6.4–11 cm)

Straddle
7.0–12 in (18–30 cm)

Stride
Walking: 16–34 in (41–86 cm)
Galloping: 3.3–7.8 ft (1.0–2.4 m)
Group length: to 6.3 ft (1.9 m)

Size (stag>hind)
Height: 4.0–5.0 ft (1.2–1.5 m)
Length: 6.5–10 ft (2.0–3.0 m)

Weight
500–1000 lb (230–450 kg)

gallop print   *walking*

ELK (Wapiti)
Cervus elaphus

Once widespread through much of Ontario, Elk are now largely absent from the province, but there are small populations in southern Manitoba. Female Elk are often seen in social herds; they like to feed in forest openings and lush meadows. Stags, who prefer to go solo, are easily recognized by their magnificent racks of antlers and haunting bugling that can be heard during rutting season (September to November).

Elk leave a neat, alternating track with large, rounded prints, often in well-worn winter paths. The hind print will sometimes double register slightly ahead of the fore print. In deeper snow, or if an Elk gallops (with its toes spread wide) the dewclaws may register. A good place to look for Elk tracks is in the soft mud by summer ponds, where Elk like to drink and sometimes splash around.

Similar Species: White-tailed Deer (p. 20) prints are generally smaller. Also compare with Moose (p. 24) prints.

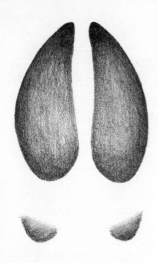

Fore and Hind Prints
Length: 4.0–7.0 in (10–18 cm)
    to 11 in (28 cm) with dewclaw
Width: 3.5–5.8 in (8.9–15 cm)

Straddle
8.5–20 in (22–51 cm)

Stride
Walking: 1.5–3.0 ft (46–91 cm)
Trot: to 4.0 ft (1.2 m)

Size (bull>cow)
Height: 5.0–6.5 ft (1.5–2.0 m)
Length: 7.0–8.5 ft (2.3–2.6 m)

Weight
600–1100 lb (270–500 kg)

walking

MOOSE
Alces alces

The impressive male Moose, largest of all the deer, has a massive rack of antlers. Moose are usually solitary, though you may see a cow and calf together. Despite their placid appearance, Moose will often charge humans if approached. Moose are found throughout most of Ontario, but not in the more developed areas of the south.

Moose move gracefully, leaving a neat, alternating track. The hind prints directly or double register on the fore prints. In summer, look for tracks in the mud beside ponds and other wet areas, where Moose especially like to feed; they are excellent swimmers. In winter they feed in willow flats and coniferous forests, leaving a distinct browseline (highline). Ripped stems and gnawed bark, 6 feet (1.8 m) or more above the ground, are signs that Moose have been around. Their long legs allow for easy movement in snow; where the print is deeper than 1.2 inches (3.0 cm), the dewclaws–which give extra support for the animal's huge weight–show far back from the main print.

Similar Species: Prints of juvenile Moose may be mistaken for those of the smaller White-tailed Deer (p. 20). Elk (p. 22), with a limited range, have smaller, rounder prints.

Fore Print
(hind print is slightly smaller)
Length: 4.5–6.0 in (11–15 cm)
Width: 4.5–5.5 in (11–14 cm)
Stride
Walking: 17–27 in (43–69 cm)
Size
Height: to 6.0 ft (1.8 m)
Weight
to 1500 lb (680 kg)

walking

HORSE
Equus caballus

This popular animal has unmistakable prints; it deserves mention because back-country use of the Horse means that you can expect its tracks to show up almost anywhere.

Unlike any other animals discussed in this book, the Horse has only one huge toe, which leaves an oval print. A distinctive feature is the 'frog' or V-shaped mark at the base of the print. When the Horse is shod, the horseshoe shows up clearly as a firm wall at the outside of the print. Not all horses will be shod, so don't expect to see this outer wall on every horse track.

A typical, leisurely horse track is an alternating walk with hind prints registering on or behind the slightly larger fore prints. Horses are capable of a range of speeds—up to a full gallop—but most recreational horseback riders prefer to walk their horses and soak up the beautiful countryside!

Similar Species: Mules have smaller prints and are rarely shod.

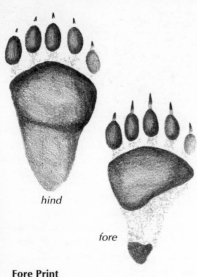

hind

fore

Fore Print
Length: 4.0–6.3 in (10–16 cm)
Width: 3.8–5.5 in (9.7–14 cm)

Hind Print
Length: 6.0–7.0 in (15–18 cm)
Width: 3.5–5.5 in (8.9–14 cm)

Straddle
9.0–15 in (23–38 cm)

Stride
Walking: 17–23 in (43–58 cm)

Size (male>female)
Height: 3.0–3.5 ft (91–110 cm)
Length: 5.0–6.0 ft (1.5–1.8 m)

Weight
200–600 lb (91–270 kg)

walking (slow)

BLACK BEAR
Ursus americanus

The Black Bear is found in forested areas of the province, but not in the populated south. Black Bears sleep deeply in the winter, so don't expect to encounter their tracks in the colder months. Finding fresh bear tracks can be a thrill, but take care, as the bear may be just around the corner. Never underestimate the potential power of a surprised bear!

A Black Bear's print is about the size of a human print, but the bear's print is wider and shows claw marks. The small inner toe rarely registers. The forefoot has a small heel pad that often shows, and the hind foot has a big heel. The bear's slow walk results in a slightly pigeon-toed double register of the hind print over the fore. At a faster pace, the hind oversteps the fore. When a bear runs, the two hind feet register in front of the forefeet in an extended cluster. Along well-worn bear paths, look for 'digs'–patches of dug-up earth–and 'bear trees,' their scratched bark and claw marks showing that these bears are capable climbers.

Similar Species: The magnificent Grizzly Bear (*Ursus arctos*) has similar prints but is now absent from most of Ontario. The Polar Bear (p. 30) has larger prints.

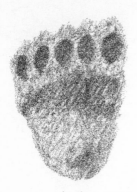

hind

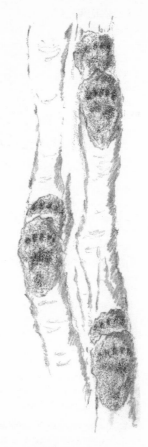

Hind Print
Length: 12–13 in (30–33 cm)
Width: to 9.0 in (23 cm)

Straddle
to 20 in (51 cm)

Size (male>female)
Height: 3.0–4.0 ft (91–120 cm)
Length: 6.5–7.5 ft (2.0–2.3 m)

Weight
600–1100 lb (270–500 kg)

walking (in snow)

POLAR BEAR
Ursus maritimus

The Polar Bear is truly the spirit of the far north. Huge and massively furred, these resilient bears roam the far northen coastal regions and the ice floes in search of carrion and seals. Few of us will have the pleasure of finding a Polar Bear's tracks, let alone the bear itself. The most likely and convenient location for an encounter with Polar Bears is Churchill, Manitoba, where many people go in order to see them.

A Polar Bear's feet are thickly haired to keep them warm. This thick fur obscures the finer details of the prints. Clear prints show five toes on each foot. The fore print is smaller than the hind print. When a Polar Bear walks, the hind print registers on or behind the fore print. The bear's feet may drag in the snow and its claws will not always register. As Polar Bear tracks are most likely to be found in snow, there is little room for confusion, because there is little else in the far north that leaves such enormous tracks.

Similar Species: The smaller Black Bear (p. 28), with similar, smaller prints, does not visit the northern coast; the Grizzly Bear (*Ursus arctos*), with longer claws, does.

fore

hind

**Fore Print
(hind print slightly smaller)**
Length: 4.0–5.5 in (10–14 cm)
Width: 2.5–5.0 in (6.4–13 cm)

Straddle
3.0–7.0 in (7.6–18 cm)

Stride
Walking: 15–32 in (38–81 cm)
Galloping: 3.0 ft (91 cm),
    leaps to 9.0 ft (2.7 m)

**Size
(female is slightly smaller)**
Height: 26–38 in (66–97 cm)
Length: 3.6–5.2 ft (1.1–1.6 m)

Weight
70–120 lb (32–55 kg)

walking   *trotting*

GRAY WOLF
(Timber Wolf)
Canis lupus

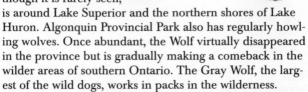

The soulful howl of the wolf epitomizes the outdoor experience. Your best bet for hearing the Wolf, though it is rarely seen, is around Lake Superior and the northern shores of Lake Huron. Algonquin Provincial Park also has regularly howling wolves. Once abundant, the Wolf virtually disappeared in the province but is gradually making a comeback in the wilder areas of southern Ontario. The Gray Wolf, the largest of the wild dogs, works in packs in the wilderness.

A wolf leaves a straight alternating track with the smaller hind print registering directly on the fore print. It has large, oval prints that each show all four claws. The lobing on the fore print heel pads is different from that of the hind ones.

If you find a wolf track in the snow, it is likely the track of several wolves, because in deep snow wolves will sensibly follow their leader, sometimes dragging their feet. When a wolf trots, notice how the hind print has a slight lead and falls to one side, giving an unbalanced appearance. Wolves and Coyotes (p. 34) gallop in the same way.

Similar Species: Domestic Dog (*Canis familiaris*) prints, rarely as large as a wolf's, fall in a haphazard track with a less direct register; the inner toes tend to spread out more.

fore

hind

Fore Print
(hind print is slightly smaller)
Length: 2.4–3.1 in (6.1–7.9 cm)
Width: 1.6–2.4 in (4.1–6.1 cm)

Straddle
4.0–7.0 in (10–18 cm)

Stride
Walking: 8.0–16 in (20–41 cm)
Galloping: 2.5–10 ft (76–300 cm)

Size
(female is slightly smaller)
Height: 23–26 in (58–66 cm)
Length: 32–40 in (81–100 cm)

Weight
20–50 lb (9.1–23 kg)

walking   *gallop group*

COYOTE (Brush Wolf, Prairie Wolf)
Canis latrans

This widespread and adaptable canine prefers to hunt rodents and larger prey in open grasslands or woodlands. It hunts on its own, with a mate or in a family pack. In southernmost Ontario, a Coyote occasionally develops an interesting cooperative relationship with a Badger, so you might find their tracks together where they have been digging for rodents.

The oval fore prints are larger than the hind prints. Note the difference between the fore heel pad and the hind heel pad, which rarely registers clearly. Claw marks are usually evident only on the two centre toes. Coyotes typically walk or trot in an alternating pattern, the walk having the wider straddle. The trail left by a Coyote is often direct, as if it knew exactly where it was going. A Coyote's tail hangs down, leaving a dragline in deep snow. When a Coyote gallops, the hind feet fall ahead of the forefeet; the faster it goes, the straighter the gallop group.

Similar Species: Domestic Dog (*Canis familiaris*) prints are not so oval and spread more and the four toes tend to register, and a dog's trail is erratic and confused. Red Fox (p. 38) prints are usually, but not always, smaller.

fore

hind

**Fore Print
(hind print slightly smaller)**
Length: 1.3–2.1 in (3.3–5.3 cm)
Width: 1.1–1.5 in (2.8–3.8 cm)

Straddle
2.0–4.0 in (5.1–10 cm)

Stride
Walking/Trotting: 7.0–12 in (18–30 cm)

Size
Height: 14 in (36 cm)
Length: 21–29 in (53–74 cm)

Weight
7.0–15 lb (3.2–6.8 kg)

walking

GRAY FOX
Urocyon cinereoargenteus

This small, shy fox is found in southernmost Ontario, especially along the shores of Lakes Ontario and Erie. Reportedly expanding its range, it can sometimes be seen in abandoned fields and on farmland bordering the woodland and shrubby areas that are its preferred habitat. Follow its track and you may be surprised to find this fox climbing a tree; it is the only fox to do so.

The fore print registers better than the hind print, on which the long, semi-retractable claws do not always show. The heel pads are often unclear–they sometimes show up just as small, round dots. This fox leaves a neat, alternating walking track; its trotting track is like a Red Fox's, and it has a gallop group like a Coyote's (p. 34).

Similar Species: The province-wide Red Fox (p. 38) has different heel pads (with a bar across them); in general, its prints are larger and less clear, its stride longer and its straddle narrower. The prints of Domestic Cats (*Felis catus*) and small Bobcats (p. 42) are similar, but lack claws and have a larger, less symmetrical heel pad.

fore

hind

**Fore Print
(hind print slightly smaller)**
Length: 2.1–3.0 in (5.3–7.6 cm)
Width: 1.6–2.3 in (4.1–5.8 cm)

Straddle
2.0–3.5 in (5.1–8.9 cm)

Stride
Walking: 12–18 in (31–46 cm)
Trotting: 14–21 in (36–53 cm)

**Size
(vixen is slightly smaller)**
Height: 14 in (36 cm)
Length: 22–25 in (56–64 cm)

Weight
7.0–15 lb (3.2–6.8 kg)

walking   *trotting*

RED FOX
Vulpes vulpes

This beautiful and notoriously cunning fox is found throughout Ontario. Very adaptable, the Red Fox prefers meadows, streambanks and other open areas, but also inhabits woodlands. It is secretive and largely nocturnal.

This fox has very hairy feet, so the finer details of its prints are obscured–only parts of the toes and heel pads show. A very significant feature unique to this fox is the horizontal or slightly curved bar across the fore heel pad.

A Red Fox leaves a distinctive straight trail of alternating prints, with the hind print direct registering on the wider fore print. When a fox trots, the paired prints show the hind print falling to one side of the fore print in typical dog-family fashion. Its gallop group is like the Coyote's. The faster the gallop, the straighter the group.

Similar Species: Domestic Dog (*Canis familiaris*) prints are of similar size but lack the bar on the heel pad, and show a shorter stride with a less direct trail. Small Coyote (p. 34) prints are similar but with a wider straddle. Gray Fox (p. 36) prints have a shorter stride and a wider straddle.

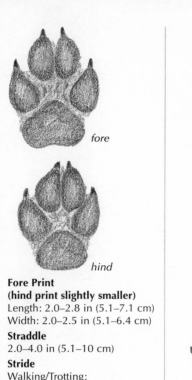

fore

hind

Fore Print
(hind print slightly smaller)
Length: 2.0–2.8 in (5.1–7.1 cm)
Width: 2.0–2.5 in (5.1–6.4 cm)

Straddle
2.0–4.0 in (5.1–10 cm)

Stride
Walking/Trotting:
   7.0–12 in (18–30 cm)

Size
Height: to 12 in (30 cm)
Length with tail:
   30–36 in (76–91 cm)

Weight
5.5–8.8 lb (2.5–4.0 kg)

loping

running

ARCTIC FOX
Alopex lagopus

This delightful fox of the far north changes from a thick and warm white winter coat to a thinner blue-brown one in summer. Usually this fox is quite a solitary animal, but it is curious and may observe you quite closely in remote areas.

Arctic Fox tracks are typical of dog-family tracks, with the hind prints slightly smaller than the fore prints. Both fore and hind prints show four toes, and the claws usually register. The toes are set closely together and register quite clearly in summer. In winter, however, when thick fur covers the pads to keep them warm, print detail is less clear. The tracks are set in a close line, leaving a neat trail. If you come across an old bear kill, look around for fox tracks, as foxes frequently follow bears in the hope that they can feed on the leftovers.

Similar Species: The Red Fox (found throughout Ontario, p. 38) has a slightly smaller print with a distinctive bar across the heel pad, though it may not be evident (because of thick winter fur).

fore

hind

Fore Print
(hind print slightly smaller)
Length: 1.8–2.5 in (4.6–6.4 cm)
Width: 1.8–2.6 in (4.6–6.6 cm)

Straddle
4.0–7.0 in (10–18 cm)

Stride
Walking: 8.0–16 in (20–41 cm)
Running: 4.0–8.0 ft (1.2–2.4 m)

Size
(female is slightly smaller)
Height: 20–22 in (51–56 cm)
Length: 25–30 in (64–76 cm)

Weight
15–35 lb (6.8–16 kg)

walking

trotting
to loping

BOBCAT (Wildcat)

Lynx rufus

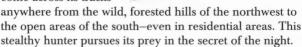

The adaptable
but seldom-seen
Bobcat is widely
distributed, except
in northernmost
Ontario. You might
come across its tracks
anywhere from the wild, forested hills of the northwest to
the open areas of the south—even in residential areas. This
stealthy hunter pursues its prey in the secret of the night.

The hind print usually registers exactly on the larger
fore print in the walking track. Fore prints especially show
asymmetry. The heel pads have two lobes to the front and
three to the rear. As a Bobcat picks up speed, its track
becomes a trot pattern made of groups of two prints, the
hind leading the fore. At even greater speeds, its track
becomes a group of four prints in a lope pattern. A Bobcat's feet leave draglines in deep snow. Unlike the trails of
wild dogs, the Bobcat's trail meanders. Half-buried scat
along the trail is a sign of a Bobcat marking its territory.

Similar Species: A juvenile Lynx (p. 44) of the northern
regions has similar prints. Large Domestic Cat (*Felis catus*)
prints may be confused with a juvenile Bobcat's prints, but
Domestic Cats have a shorter stride and a narrower
straddle and do not wander far from home, especially in
winter. Note: dog, coyote and fox prints show claw marks;
the fronts of their footpads are singly lobed.

fore

hind

Fore Print
(hind print slightly smaller)
Length: 3.5–4.5 in (8.9–11 cm)
Width: 3.5–4.8 in (8.9–12 cm)

Straddle
6.0–9.0 in (15–23 cm)

Stride
Walking: 12–28 in (30–71 cm)

Size
Length: 29–36 in (74–91 cm)

Weight
15–30 lb (6.8–14 kg)

walking

LYNX (Canada Lynx)
Lynx lynx

This large cat eludes humans by living in dense forests. It is absent from southernmost Ontario. The Lynx is sensitive to human interference and is therefore more abundant in remote regions. With its huge feet and relatively light body weight, the Lynx stays on top of the snow while in pursuit of its main prey, the Snowshoe Hare.

This cautious walker leaves an alternating track with neat direct registers of the hind print over the fore print. Thick fur obscures the prints' characteristics so that they often appear as big, round depressions with no detail. In deeper snow, the print may be extended by 'handles' off to the rear, but the Lynx rarely drags its feet. However deep the snow, this cat sinks no more than 8 inches (20 cm). Its curious nature results in a meandering trail, which may lead you to a partially buried cache of food.

Similar Species: The Bobcat (p. 42) has smaller prints and is more widely distributed in southern regions than the Lynx, which ranges farther north and is more likely to bound than run.

45

fore

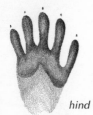

hind

Fore Print
Length: 2.0–3.0 in (5.1–7.6 cm)
Width: 1.8–2.5 in (4.6–6.4 cm)

Hind Print
Length: 2.4–3.8 in (6.1–9.7 cm)
Width: 2.0–2.5 in (5.1–6.4 cm)

Straddle
3.3–6.0 in (8.4–15 cm)

Stride
Walking/Running:
   7.0–20 in (18–51 cm)

Size (females slightly smaller)
Length: 24–37 in (61–94 cm)

Weight
11–35 lb (5.0–16 kg)

walking   *running group*

RACCOON
Procyon lotor

The inquisitive Raccoon is adored by some for its distinctive face mask, yet disliked for its boundless curiosity—often demonstrated with garbage cans. Raccoons are common throughout most of the province. Look for their tracks around bodies of water, in a diversity of habitats. Raccoons usually den up for the colder months of winter.

The Raccoon's unusual print looks like a human handprint, showing five well-formed toes. The small claws appear as dots. Its highly dexterous forefeet rarely leave heel prints, but its hind prints, which are generally much clearer, do register heels. The Raccoon's peculiar walking track shows a left fore print next to a right hind print and vice versa. The fore print may be ahead of the hind print. Raccoons occasionally run, leaving clusters in which the two hind prints fall ahead of the fore prints. A Raccoon's trail may lead you up a tree, where Raccoons like to rest.

Similar Species: The trail of the Fisher (p. 52), which prefers deeper forests, has similar prints but shows different gaits. Unclear Opossum (p. 66) prints may look similar, but Opossums drag their tails.

fore

hind

Fore Print
Length: 2.5–3.5 in (6.4–8.9 cm)
Width: 2.0–3.0 in (5.1–7.6 cm)

Hind Print
Length: 3.0–4.0 in (7.6–10 cm)
Width: 2.3–3.3 in (5.8–8.4 cm)

Straddle
4.0–9.0 in (10–23 cm)

Stride
Walking/Running: 6.0–23 in (15–58 cm)

Size
(female is slightly smaller)
Length with tail: 3.0–4.3 ft (91–130 cm)

Weight
10–25 lb (4.5–11 kg)

running (fast)

RIVER OTTER
Lutra canadensis

No animal knows how to have more fun than a
River Otter. If you are lucky enough to watch one at play,
you will not soon forget the experience. The River Otter
is found throughout Ontario, near streams and other
waterbodies. An otter in the forest is usually travelling
between waterbodies. Within an otter's home territory, you
are sure to find a wealth of evidence along the riverbanks.

In soft mud, the River Otter's five-toed feet show
evidence of webbing, the hind prints more than the fore
prints. The inner toes are set slightly apart. The heel may
register, lengthening the print. Otter tracks are very
variable: they show the typical two-print bounding of the
weasel family, and, with faster runs, groups of four and
three prints. The thick, heavy tail often leaves a mark
over the prints. Otters love to slide in snow, often down
riverbanks, leaving troughs nearly 12 inches (30 cm) wide.
In summer they roll and slide on grass and mud.

Similar Species: Fisher (p. 52) prints are similar, but
Fishers prefer forests, do not have webbed feet and do
not leave conspicuous tail draglines.

fore left

hind left

Fore and Hind Prints
Length: 1.3–2.0 in (3.3–5.1 cm)
Width: 1.3–1.8 in (3.3–4.6 cm)

Straddle
2.1–3.5 in (5.3–8.9 cm)

Stride
Walking/Running: 8.0–35 in (20–89 cm)

Size
(female is slightly smaller)
Length with tail: 19–28 in (48–71 cm)

Weight
1.5–3.5 lb (0.7–1.6 kg)

running

MINK
Mustela vison

The lustrous Mink prefers watery habitats surrounded by brush or forest and it is widespread throughout the province. At home as much on land as in water, this nocturnal hunter can be an exciting animal to track. Like the River Otter, the Mink slides in snow, leaving a trough carved out by its body for an observant tracker to spot.

The fore print of the Mink shows five (sometimes four) toes with five loosely connected palm pads in an arc, but the hind print shows only four palm pads. The heel pad rarely shows on the fore print but may register on the hind print, lengthening it. The Mink prefers the typical weasel-family bounding pattern of double prints, consistently spaced and slightly angled. Nevertheless, its tracks show much diversity in gait and may also appear as an alternating walk, or as a run with three- and four-print groups, as illustrated with the River Otter (p. 48).

Similar Species: A small Marten (p. 54) print may be mistaken for a large Mink print, but Martens stay away from water, and their tracks do not show such consistent double-print bounding patterns.

51

 fore

Fore Print
Length: 2.1–3.9 in
    (5.3–9.9 cm)
Width: 2.1–3.3 in
    (5.3–8.4 cm)

Hind Print
Length: 2.1–3.0 in (5.3–7.6 cm)
Width: 2.0–3.0 in (5.1–7.6 cm)

Straddle
3.0–7.0 in (7.6–18 cm)

Stride
Walking: 7.0–14 in (18–36 cm)
Running: 1.0–4.2 ft (30–130 cm)

Size (male>female)
Length with tail:
    33–41 in (84–100 cm)

Weight
3.0–12 lb (1.4–5.4 kg)

walking

running

FISHER (Black Cat)

Martes pennanti

This agile hunter is comfortable both on the ground and in the trees of the mixed hardwood forests along the northern shores of Lakes Superior and Huron but is absent from the south and east of the province. Its speed and eager hunting antics make for exciting tracking–it races up trees and along the ground in its quest for squirrels. A Fisher's tracks may lead to the plucked remains of an unfortunate Porcupine; it kills and eats Porcupines (p. 76) regularly, one of the very few predators to do so.

The Fisher occasionally leaves a direct-registering alternating walking track. However, it usually prefers to run in typical weasel fashion, leaving angled pairs of prints that represent the direct register of hind print over fore print. While five toes may register, the small inner toe frequently does not. Only the forefoot has a small heel pad that can show up in the print. Tracks often vary within a short distance, and are not associated with water–the name 'Fisher' is a misnomer!

Similar Species: Small female Fisher tracks may be confused with male Marten (p. 54) tracks, but a Marten weighs less and leaves shallower prints. When only four toes register, Fisher prints may look like a Bobcat's (p. 42).

fore

Fore and Hind Prints
Length: 1.8–2.5 in
    (4.6–6.4 cm)
Width: 1.5–2.8 in
    (3.8–7.1 cm)

Straddle
2.5–4.0 in (6.4–10 cm)

Stride
Walking: 4.0–9.0 in
    (10–23 cm)
Running: 9.0–46 in
    (23–120 cm)

Size (male>female)
Length with tail:
    21–27 in (53–69 cm)

Weight
1.5–2.8 lb (0.7–1.3 kg)

walking   *running*

MARTEN (Pine Marten, American Sable)
Martes americana

This aggressive
predator is found in
the coniferous forests of much of Ontario,
but not in the more populated south.

A Marten seldom leaves a clear print: the heel pad
is very undeveloped and, in winter, the hairiness of the
feet often obscures all pad detail from the print, especially
the detail of the poorly developed palm pads. Prints often
show only four toes, when the inner toe fails to register.

In the Marten's alternating walking pattern, the hind
print registers on the fore print. In a bounding track, the
hind prints fall on the fore prints, in slightly angled groups
of two, in the typical weasel-family pattern. Gallop groups
may appear as clusters of three or four prints. Follow the
criss-crossing tracks and you may find that the Marten has
scrambled up a tree in pursuit of a Red Squirrel (p. 88);
look for the sitzmark where it has jumped down again.

Similar Species: Female Fisher (p. 52) prints overlap in
size with large male Marten prints, but Fishers leave clearer
tracks. Male Mink (p. 50) prints overlap in size with small
female Marten prints, but Mink do not climb trees and,
unlike Martens, are often found near water.

all weasels, left

LEAST WEASEL
Fore and Hind Prints
Length: 0.5–0.8 in (1.3–2.0 cm)
Width: 0.4–0.5 in (1.0–1.3 cm)
Straddle
0.8–1.5 in (2.0–3.8 cm)
Stride
Bounding: 5.0–20 in (13–51 cm)
Size
(male>female)
Length with tail: 6.5–9.0 in (17–23 cm)
Weight
1.3–2.3 oz (37–65 g)

Least Weasel
(bounding)

WEASELS
Mustela spp.

Weasels are active hunters, with an avid appetite for rodents. Following their tracks can reveal much about these nimble creatures' activities. Although weasels are active all year, their tracks are most evident in winter, when they frequently burrow into the snow or, in pursuit of their prey, enter an existing rodent hole. From time to time, weasel tracks may lead you up a tree; weasels have also been known to take to water.

The typical trail of the weasel is a bounding pattern of paired prints. Because of a weasel's light weight and small, hairy feet, the pad detail is often obscured, especially in snow. Even with clear tracks, the inner (fifth) toe rarely registers. Successful identification of which weasel made the track can be troublesome. Pay close attention to the straddle and stride, but note that small females of a larger species and large males of a smaller species may have similar tracks. Clues can be gained from the habit displayed in jumping patterns. Also check distribution and habitat.

LEAST WEASEL
Mustela nivalis

The Least Weasel is the smallest weasel, with the least defined track. This weasel is found throughout the province. Its tracks may be found around wetlands and in open woodlands and fields.

Similar Species: A large male's tracks may resemble a small female Short-tailed Weasel's, but the latter does not frequent wet areas, preferring upland areas and woodlands.

SHORT-TAILED WEASEL
Fore and Hind Prints
Length: 0.8–1.3 in (2.0–3.3 cm)
Width: 0.5–0.6 in (1.3–1.5 cm)

Straddle
1.0–2.1 in (2.5–5.3 cm)

Stride
Bounding: 9.0–35 in (23–89 cm)

Size (male>female)
Length with tail:
8.0–14 in (20–36 cm)

Weight
1.0–6.0 oz (28–170 g)

LONG-TAILED WEASEL
Fore and Hind Prints
Length: 1.1–1.8 in (2.8–4.6 cm)
Width: 0.8–1.0 in (2.0–2.5 cm)

Straddle
1.8–2.8 in (4.6–7.1 cm)

Stride
Bounding: 9.5–43 in (24–110 cm)

Size (male>female)
Length with tail:
12–22 in (30–56 cm)

Weight
3.0–12 oz (85–340 g)

Short-tailed Weasel (bounding)   *Long-tailed Weasel (bounding)*

SHORT-TAILED WEASEL (Ermine)
Mustela erminea

This weasel is larger than the Least Weasel, but smaller than the Long-tailed Weasel. Like the Least Weasel, the Short-tailed Weasel is found across the province. It prefers woodlands and meadows, but is not noted to favour the densest coniferous forests or wetlands. Its bounding track may show short strides alternating with long ones.

Similar Species: A small female's tracks may resemble a large male Least Weasel's. A large male's tracks may resemble a small female Long-tailed Weasel's.

LONG-TAILED WEASEL
Mustela frenata

Long-tailed Weasel

The Long-tailed Weasel, the largest of the three, is widely distributed in the south and east of Ontario, and along the border with Minnesota, but is absent from the northern shores of Lake Superior and from the far north. Its tracks, larger than for other weasels, have an inconsistent, irregular stride–sometimes short, sometimes long.

Similar Species: A small female may leave tracks the same size as a large male Short-tailed Weasel's.

fore

hind

**Fore Print
(hind print slightly shorter)**
Length: 2.5–3.0 in (6.4–7.6 cm)
Width: 2.3–2.8 in (5.8–7.1 cm)

Straddle
4.0–7.0 in (10–18 cm)

Stride
Walking: 6.0–12 in (15–30 cm)

Size
Length: 21–35 in (53–89 cm)

Weight
13–25 lb (5.9–11 kg)

walking

BADGER
Taxidea taxus

The squat shape and unmistakable face of this bold animal are features suitable for the open grasslands and meadows. It is found only in the southernmost part of Ontario and possibly along the western shores of Lake Superior. Thick shoulders and forelegs, coupled with long claws, make for a powerful digging animal. Unlike most other weasel-family members, the Badger likes to den up in a hole during the really cold months of winter in more northerly regions, so look for its tracks in spring and fall snow.

All five toes on each foot register. A Badger's long claws are evident in the pigeon-toed track that it leaves as it waddles along, although the claws on the hind feet are not as long as those on the forefeet. When a Badger walks, the alternating track is a double register, with the hind print sometimes falling just behind the fore print and sometimes slightly ahead. The Badger's wide, low body will often plough through deeper snow and obscure the print detail.

Similar Species: In snow, Porcupine (p. 76) tracks may be similar but show draglines made by its tail and quills and will likely lead up a tree, not to a hole.

fore

hind

Fore Print
Length: 1.5–2.2 in
 (3.8–5.6 cm)
Width: 1.0–1.5 in (2.5–3.8 cm)

Hind Print
Length: 1.5–2.5 in
 (3.8–6.4 cm)
Width: 1.0–1.5 in (2.6–3.8 cm)

Straddle
2.8–4.5 in (7.1–11 cm)

Stride
Walking/Running:
 2.5–8.0 in (6.4–20 cm)

Size
Length with tail:
 20–32 in (51–81 cm)

Weight
6.0–14 lb (2.7–6.4 kg)

walking (fast)

running

STRIPED SKUNK
Mephitis mephitis

This striking skunk has a notorious reputation for its vile smell; the lingering odour is often the best sign of its presence. Widespread throughout all but the extreme north of Ontario, the Striped Skunk enjoys a diverse range of habitats, especially open woodlands and brushy areas. It dens up in winter, coming out on warmer days and in the spring.

Both fore and hind feet have five toes. The long claws on the forefeet often register. Smooth palm pads and small heel pads leave surprisingly small prints. Skunks mostly walk–with such a potent smell for their defence, and those memorable black and white stripes, they rarely need to run. Unlike other weasel-family members, skunks rarely show any consistent pattern in their tracks, but an alternating walking pattern may be evident. The greater the skunk's speed, the more its hind print oversteps the fore. Should a skunk need to run, its track is a closely set pattern of clumsy four-print groups. In snow, skunks drag their feet.

Similar Species: Skunk tracks are quite distinctive and recognizable. Eastern Spotted Skunks (p. 64), with smaller prints, may be a rare find along the border with Minnesota.

fore

hind

Fore Print
Length: 1.0–1.3 in (2.5–3.3 cm)
Width: 0.9–1.1 in (2.3–2.8 cm)

Hind Print
Length: 1.2–1.5 in (3.0–3.8 cm)
Width: 0.9–1.1 in (2.3–2.8 cm)

Straddle
2.0–3.0 in (5.1–7.6 cm)

Stride
Walking: 1.5–3.0 in
  (3.8–7.6 cm)
Jumping: 6.0–12 in (15–30 cm)

Size
Length: 13–25 in (33–64 cm)

Weight
0.6–2.2 lb (0.3–1.0 kg)

walking

running

EASTERN SPOTTED SKUNK

Spilogale putorius

This beauti-
fully marked
skunk is smaller
than its striped
cousin. The Eastern
Spotted Skunk,
which is found
only near the border with Minnesota, enjoys diverse habitats, such as scrubland, forests and farmland. This skunk is
a rare sight, since it is largely nocturnal. It dens up to avoid
the coldest months, coming out only on warmer days.

This skunk leaves a very haphazard trail as it forages for
food on the ground. Occasionally, with ease, it climbs trees.
Long claws on the forefeet often register, and the palm and
heel may show some defined pads. Although this skunk
rarely runs, when it does so it may bound along, leaving
groups of four prints, hind ahead of fore. This skunk sprays
only when truly provoked, so its powerful odour is less
frequently detected than that of the Striped Skunk.

Similar Species: The bigger Striped Skunk (p. 62), with a
widespread distribution throughout the province, has larger
prints and less-scattered tracks with a shorter running stride
(or it jumps); it does not climb trees.

fore

hind

Fore Print
Length: 2.0–2.3 in (5.1–5.8 cm)
Width: 2.0–2.3 in (5.1–5.8 cm)

Hind Print
Length: 2.5–3.0 in (6.4–7.6 cm)
Width: 2.0–3.0 in (5.1–7.6 cm)

Straddle
4.0–5.0 in (10–13 cm)

Stride
5.0–11 in (13–28 cm)

Size
Length: 2.0–2.5 ft (61–76 cm)

Weight
9.0–13 lb (4.1–5.9 kg)

walking   *running*

OPOSSUM
Didelphis virginiana

This slow-moving nocturnal marsupial inhabits southern Ontario; the bitter cold of the north likely prevents it from extending its range. While the Opossum is found in many types of habitat, it shows a preference for open woodland or brushland around waterbodies. It is quite tolerant of residential areas. Opossum tracks can often be seen in mud near the water and in snow during the warmer months of winter–Opossums tend to den up during severe weather.

Opossums are excellent climbers, so do not be surprised if their tracks lead to a tree. They have two walking habits: the common alternating pattern, with the hind prints registering on the fore prints, and a Raccoon-like paired-print pattern, with the hind print next to the opposing fore print. The long, very distinctive, inward-pointing thumb of the hind foot does not show a claw mark. In snow, the drag line from the long, naked tail may be stained with blood–this thinly haired animal frequently suffers frostbite.

Similar Species: Prints in which the distinctive thumbs don't show may be mistaken for a Raccoon's (p. 46).

hind

fore

Fore Print
Length: 2.5–4.0 in (6.4–10 cm)
Width: 1.3–1.7 in (3.3–4.3 cm)

Hind Print
Length: 3.5–6.5 in (8.9–17 cm)
Width: 1.5–2.5 in (3.8–6.4 cm)

Straddle
4.5–7.0 in (11–18 cm)

Stride
Hopping: 1.0–10 ft (30–300 cm)

Size
Length: 23–25 in (58–64 cm)

Weight
5.0–9.0 lb (2.3–4.1 kg)

hopping

WHITE-TAILED JACKRABBIT

Lepus townsendii

This hare frequents the open country of the extreme western parts of the province. An athletic animal, the White-tailed Jackrabbit can reach the impressive speed of 45 mph (72 km/h). Because of its nocturnal and solitary habits and its wariness of predators, it is infrequently seen.

Both fore and hind prints show four toes; the hind foot may often register a long heel. When it hops, this hare creates print groups in a triangular pattern; as it speeds up, these print groups spread out considerably. Following the tracks could lead you to its 'form'—a depression where the hare rests—or an urgent zigzag pattern that indicates where the hare fled from danger. With its strong hind legs, it is capable of leaping up to 20 feet (6.1 m) to avoid pursuers.

Similar Species: The Snowshoe Hare (p. 70) spreads its hind toes more, takes shorter leaps and requires dense cover. Coyote (p. 34) prints resemble heel-less jackrabbit prints, but the gait is very different. The Eastern Cottontail (p. 74) has much smaller print clusters and shorter strides. The European Hare (p. 72) has a different range.

fore

hind

Fore Print
Length: 2.0–3.0 in (5.1–7.6 cm)
Width: 1.5–2.0 in (3.8–5.1 cm)

Hind Print
Length: 4.0–6.0 in (10–15 cm)
Width: 2.0–3.5 in (5.1–8.9 cm)

Straddle
6.0–8.0 in (15–20 cm)

Stride
Hopping: 0.8–4.2 ft (24–130 cm)

Size
Length: 12–21 in (30–53 cm)

Weight
2.0–4.0 lb (0.9–1.8 kg)

hopping

SNOWSHOE HARE
(Varying Hare)
Lepus americanus

This hare is well known for its colour change, from summer brown to winter white, and for its huge hind feet, which enable it to 'float' on the surface of snow. Widespread in Ontario, it frequents brushy areas in forests, as these areas provide good cover from Lynx and Coyotes, its most likely predators. Well-worn runways are used as escape runs. Hares are most active at night.

As with rabbits and other hares, the Snowshoe Hare usually leaves a hopping track, with groups of four prints in a triangular pattern; they can be quite long if the hare is running quickly. A hare track's most distinctive feature is that the hind print is much larger than the fore print. In winter, heavy fur thickens the toes of a Snowshoe Hare's hind feet, which splay out when it runs, distributing its weight when it runs on snow. If you are lucky, you might even come across a resting hare, since they do not live in burrows. Signs of this hare's presence include twigs and stems that have been neatly severed at a 45° angle.

Similar Species: Eastern Cottontail (p. 74) prints are similar, but smaller. The White-tailed Jackrabbit (p. 68) and the European Hare (p. 72) spread their hind-foot toes less.

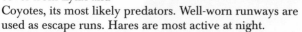

71

fore

hind

Fore Print
Length: 2.0–4.0 in (5.1–10 cm)
Width: 1.3 in (3.3 cm)

Hind Print
Length: 5.0 in (13 cm)
　　9.0 in (23 cm) with heel
Width: 2.5 in (6.4 cm)

Straddle
8.0 in (20 cm)

Stride
Hopping: 1.5–3.0 ft (46–91 cm)
Bounding: to 12 ft (3.7 m)

Size
Length: 25–30 in (64–76 cm)

Weight
7.0–12 lb (3.2–5.4 kg)

hopping

EUROPEAN HARE
Lepus europaeus

Much larger than the Eastern Cottontail and the Snowshoe Hare, this sturdy coloniser of open fields was introduced to North America over one hundred years ago. It can now be found in much of central and southern Ontario and along the shores of Lake Superior. It looks much like the White-tailed Jackrabbit, but has a black spot on its tail. In winter the European Hare remains brown–unlike the Snowshoe Hare and White-tailed Jackrabbit, which turn white.

The typical track of this hare is much like those of other rabbits and hares: a cluster of prints falling in an elongated triangle. The forefeet register first (usually in a slightly diagonal line), leaving small, roundish prints that show four toes. The two much larger hind prints then fall (usually side by side) ahead of the fore prints. The hind prints are much longer when the whole heel registers, as when the hare stops for a moment.

Similar Species: The Eastern Cottontail (p. 74) has similar but much smaller tracks. Snowshoe Hare (p. 70) prints can be distinguished by their toes, which spread out more in snow. The White-tailed Jackrabbit (p. 68) has prints of similar size, but it is found only in the extreme west.

fore

hind

Fore Print
Length: 1.0–1.5 in (2.5–3.8 cm)
Width: 0.8–1.3 in (2.0–3.3 cm)

Hind Print
Length: 3.0–3.5 in (7.6–8.9 cm)
Width: 1.0–1.5 in (2.5–3.8 cm)

Straddle
4.0–5.0 in (10–13 cm)

Stride
Hopping: 0.6–3.0 ft (18–91 cm)

Size
Length: 12–17 in (30–43 cm)

Weight
1.3–3.0 lb (0.6–1.4 kg)

hopping

EASTERN COTTONTAIL
Sylvilagus floridanus

This abundant rabbit is found in southern Ontario and up into Quebec. Preferring brushy areas in grassland and cultivated areas, it might be found in dense vegetation, hiding from predators such as the Bobcat and the Coyote. The Eastern Cottontail is largely nocturnal, but can be seen out and about at dawn, at dusk and on darker days.

As with other rabbits and hares, its most common track is a triangular grouping of four prints, with the larger hind prints falling ahead of the fore prints. The two fore prints may overlap. The hairy toes will obscure any pad detail that you might hope for. The hind prints can often appear rather pointed. If you follow its tracks, you could be startled if the rabbit flies out from its 'form' (a depression in the snow or the ground where it rests).

Similar Species: The White-tailed Jackrabbit (p. 68) of the far west leaves much larger print clusters and has longer strides, as does the European Hare (p. 72). The Snowshoe Hare (p. 70) also has larger prints, especially the hind ones. Squirrel (pp. 86–93) tracks show a similar pattern, but fore prints are more consistently side by side.

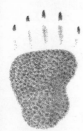

fore

hind

Fore Print
Length: 2.3–3.3 in (5.8–8.4 cm)
Width: 1.3–1.9 in (3.3–4.8 cm)

Hind Print
Length: 2.8–3.9 in (7.1–9.9 cm)
Width: 1.5–2.0 in (3.8–5.1 cm)

Straddle
5.5–9.0 in (14–23 cm)

Stride
Walking: 5.0–10 in (13–25 cm)

Size
Length with tail:
    2.2–3.4 ft (67–100 cm)

Weight
10–28 lb (4.5–13 kg)

walking

PORCUPINE
Erethizon dorsatum

This notorious
and easily recognized
rodent rarely runs, because
its many long quills are a
formidable defence. Widespread
throughout Ontario, it prefers conifer-
ous or mixed forests, but also inhabits more open areas.

The most common Porcupine track is an alternating
walking pattern, with the longer hind print registering on or
slightly ahead of the fore print. Look for long claw marks
on both prints. The fore print shows four toes and the hind
print five. On clear prints, the unusual pebbly surface of the
solid heel pads may show. However, a Porcupine's pigeon-
toed footprints are often obscured by scratches from its
heavy, spiny tail. A Porcupine's waddling gait shows in its
track. In deeper snow, this squat animal drags its feet, and it
may leave a trough with its body. A Porcupine's trail might
lead you to a tree, where these animals spend much of their
time feeding—look for chewed bark or nipped buds lying on
the forest floor.

Similar Species: Badgers (p. 60) have pigeon-toed prints,
but they don't drag their tails or climb trees.

fore

hind

Fore Print
Length: 2.5–4.0 in (6.4–10 cm)
Width: 2.0–3.5 in (5.1–8.9 cm)

Hind Print
Length: 5.0–7.0 in (13–18 cm)
Width: 3.3–5.3 in (8.4–13 cm)

Straddle
6.0–11 in (15–28 cm)

Stride
Walking: 3.0–6.5 in (7.6–17 cm)

Size
Length with tail: 3.0–3.9 ft (91–120 cm)

Weight
28–75 lb (13–34 kg)

walking

BEAVER
Castor canadensis

Few animals leave as many signs of their presence as the widespread Beaver, the largest North American rodent and a common sight around water. Signs of Beaver activity include their dams and domed lodges made of sticks and branches, and the stumps of felled trees–trunks gnawed clean of bark bear marks of the Beaver's huge incisors. Beavers also make scent mounds marked with castoreum, a strong-smelling yellowish fluid that they produce.

A Beaver's tracks are often obscured by its thick, scaly tail, or by the branches that it drags about for construction and food, but if you do find tracks where individual prints are evident, look for webbing on the large hind prints. Broad toenails are another feature of the hind print, but the inner second toenail usually does not show. Although Beavers have five toes on each foot, it is rare for all of them to register. The Beaver's track may be in an irregular alternating sequence, often with a double register. Beavers often use the same path, resulting in a well-worn trail.

Similar Species: Little confusion arises, because Beavers leave so many distinctive signs of their presence.

fore

hind

Fore Print
Length: 1.1–1.5 in (2.8–3.8 cm)
Width: 1.1–1.5 in (2.8–3.8 cm)

Hind Print
Length: 1.6–3.2 in (4.1–8.1 cm)
Width: 1.5–2.1 in (3.8–5.3 cm)

Straddle
3.0–5.0 in (7.6–13 cm)

Stride
Walking: 3.0–5.0 in (7.6–13 cm)
Running: to 1.0 ft (30 cm)

Size
Length with tail: 16–25 in (41–64 cm)

Weight
2.0–4.0 lb (0.9–1.8 kg)

walking

MUSKRAT
Ondatra zibethica

Like the Beaver, this rodent is found throughout the province, wherever there is water. Beavers (p. 78) are very tolerant of Muskrats and even allow them to live in parts of their lodges. Muskrats are active all year, and they leave plenty of signs of their presence. They dig an extensive network of burrows, often undermining the riverbank, so do not be surprised if you suddenly fall into a hidden hole! Other signs of this rodent are their small lodges in the water, and the beds of vegetation on which they rest, sun and feed during the summer.

The reduced inner toe of the five on each forefoot rarely registers in the print, but the hind print shows five well-formed toes. The prints are usually in an alternating track, with the hind print just behind or slightly overlapping the fore print. In snow, a Muskrat's feet drag a lot, and its tail leaves a sweeping dragline.

Similar Species: Few animals share this water-loving rodent's habits.

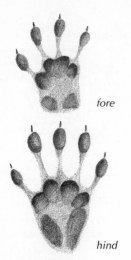

fore

hind

Fore and Hind Prints
Length: 1.8–2.8 in
    (4.6–7.1 cm)
Width: 1.0–2.0 in (2.5–5.1 cm)

Straddle
3.3–6.0 in (8.4–15 cm)

Stride
Walking: 2.0–6.0 in
    (5.1–15 cm)
Running: 6.0–14 in (15–36 cm)

Size (male>female)
Length with tail:
    20–25 in (51–64 cm)

Weight
5.5–12 lb (2.5–5.4 kg)

walking   *running*

WOODCHUCK
(Whistle Pig, Groundhog, Marmot)
Marmota monax

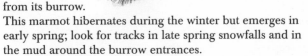

This robust member of the squirrel family is a common sight in open woodlands and adjacent open areas throughout the province. Always on the watch for predators, but not too troubled by humans, the Woodchuck will never wander too far from its burrow. This marmot hibernates during the winter but emerges in early spring; look for tracks in late spring snowfalls and in the mud around the burrow entrances.

A Woodchuck's fore print shows four toes, three palm pads and two heel pads, although the heel pads are not always evident. The hind print shows five toes, four palm pads and two poorly registering heel pads. Woodchucks usually walk, leaving an alternating pattern with the hind print registering over the fore print. When a Woodchuck runs to escape danger, it makes groups of four prints, hind ahead of fore.

Similar Species: Small Raccoon (p. 46) running tracks are similar, but have five-toed fore prints.

fore

hind

Fore Print
Length: 0.7–0.8 in (1.8–2.0 cm)
Width: 0.5 in (1.3 cm)

Hind Print
Length: 1.0–1.3 in (2.5–3.3 cm)
Width: 0.8–1.0 in (2.0–2.5 cm)

Straddle
3.0 in (7.6 cm)

Stride
Walking: 1.5–3.5 in (3.8–8.9 cm)
Jumping: 5.0–12 in (13–30 cm)

Size
Length with tail:
   13–19 in (33–48 cm)

Weight
7.0–18 oz (200–510 g)

walking

NORWAY RAT

Rattus norvegicus

This despised rat is widespread almost anywhere humans have decided to build homes. Although it is not entirely dependent on people, it rarely lives in the wild. It is active both day and night.

The Norway Rat commonly leaves an alternating walking pattern, with the larger hind print registering close to or on the fore; the hind heel does not show. The fore print shows four toes, while the hind print shows five. When it runs, this colonial rat leaves groups of four prints with the diagonally placed fore prints registering behind the hind prints. In snow, the rat's tail often leaves a drag-line. Since rats live in groups, you may find that there are many tracks close together, often leading to their 2-inch (5.1 cm) wide burrows in the ground.

Similar Species: Woodrat (*Neotoma* spp.) tracks may be similar, but woodrats rarely associate with human activity, except in abandoned buildings.

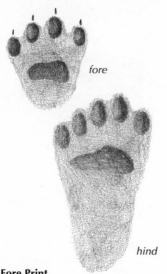

fore

hind

Fore Print
Length: 0.5–0.8 in (1.3–2.0 cm)
Width: 0.5 in (1.3 cm)

Hind Print
Length: 1.3–1.8 in (3.3–4.6 cm)
Width: 0.8 in (2.0 cm)

Straddle
3.0–3.8 in (7.6–9.7 cm)

Stride
Running: 11–29 in (28–74 cm)

Size
Length with tail: 9.0–12 in (23–30 cm)

Weight
4.0–6.5 oz (110–180 g)

sitzmark into running

NORTHERN FLYING SQUIRREL

Glaucomys sabrinus

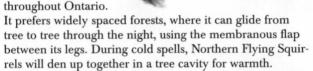

This soft-furred brown acrobat is found in coniferous and mixed forests throughout Ontario. It prefers widely spaced forests, where it can glide from tree to tree through the night, using the membranous flap between its legs. During cold spells, Northern Flying Squirrels will den up together in a tree cavity for warmth.

Because of its gliding ability, this squirrel does not leave as many tracks as the Red Squirrel and it is very difficult to find any evidence of its presence in summer. In winter, however, you might come across a sitzmark—the distinctive shape that a squirrel makes when it lands on the ground—and a short bounding track, made as the squirrel rushed off to the nearest tree or to do some quick foraging. The bounding tracks left in snow are typical of squirrels and other rodents, with the hind prints falling ahead of the fore prints, which usually register side by side.

Similar Species: The smaller, greyer Southern Flying Squirrel (*Glaucomys volans*) leaves smaller prints and is found only in the south. Red Squirrels (p. 88) usually have larger prints and never leave sitzmarks, but in deep snow their tracks resemble a Northern Flying Squirrel's. Chipmunks (p. 94) have a smaller straddle and smaller prints.

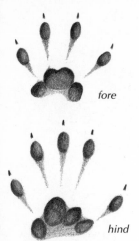

fore

hind

running

running (in deep snow)

Fore Print
Length: 0.8–1.5 in (2.0–3.8 cm)
Width: 0.5–1.0 in (1.3–2.5 cm)

Hind Print
Length: 1.5–2.3 in (3.8–5.8 cm)
Width: 0.8–1.3 in (2.0–3.3 cm)

Straddle
3.0–4.5 in (7.6–11 cm)

Stride
Running: 8.0–30 in (20–76 cm)

Size
Length with tail:
    9.0–15 in (23–38 cm)

Weight
2.0–9.0 oz (57–260 g)

RED SQUIRREL
(Pine Squirrel, Chickaree)
Tamiasciurus hudsonicus

When you enter its territory, a Red Squirrel greets you with a loud, chattering call. Another obvious sign of this province-wide forest dweller is the large middens—piles of cone scales and cores—at the bases of trees that indicate its favourite feeding sites.

Active year-round in their small territories, Red Squirrels make an abundance of tracks that lead from tree to tree or down a burrow. These energetic animals mostly run, with a gait that leaves groups of four prints, hind prints falling ahead of fore prints, which are often side by side. Four toes show on each fore print, and five on each hind print. The heels often do not register when a Red Squirrel runs. In deeper snow, the prints merge to form pairs of diamond-shaped tracks.

Similar Species: Fox Squirrel (p. 92) and Eastern Gray Squirrel (p. 90) prints are larger. A rabbit's or hare's (pp. 68–75) fore prints rarely register side by side when it runs, and it leaves a longer group of four. Chipmunk (p. 92) tracks and Northern Flying Squirrel (p. 86) tracks, in similar patterns, have smaller straddles and smaller prints.

fore

hind

Fore Print
Length: 1.0–1.8 in (2.5–4.6 cm)
Width: 1.0 in (2.5 cm)

Hind Print
Length: 2.3–3.0 in (5.8–7.6 cm)
Width: 1.1–1.5 in (2.8–3.8 cm)

Straddle
3.8–6.0 in (9.7–15 cm)

Stride
Running: 0.7–3.0 ft (21–91 cm)

Size
Length with tail: 17–20 in (43–51 cm)

Weight
14–25 oz (400–710 g)

running

EASTERN GRAY SQUIRREL

Sciurus carolinensis

This large and familiar squirrel can be a common sight in southern Ontario's deciduous and mixed forests, even in urban areas. Active all year, the Eastern Gray Squirrel can leave a wealth of evidence, especially in winter, as it scurries about digging up nuts that it buried during the previous fall.

The Eastern Gray Squirrel leaves a typical squirrel track when it runs or bounds. The hind prints fall slightly ahead of the fore prints. A clear fore print shows four toes with sharp claws, four fused palm pads and two heel pads. The hind print shows five toes and four palm pads; if the full length of the heel registers, it shows two small heel pads.

Similar Species: Fox Squirrel (p. 92) prints are as big or larger, but the animals have different ranges. Red Squirrel (p. 88) prints are smaller. A rabbit or hare (pp. 68–75) makes a longer print pattern; its fore prints rarely register side by side when it runs. Chipmunk (p. 94) tracks and Northern Flying Squirrel (p. 86) tracks, although in similar patterns, have smaller straddles and smaller prints.

fore

hind

Fore Print
Length: 1.0–1.9 in (2.5–4.5 cm)
Width: 1.0–1.7 in (2.5–4.3 cm)

Hind Print
Length: 2.0–3.3 in (5.1–7.6 cm)
Width: 1.5–1.9 in (3.8–4.8 cm)

Straddle
4.0–6.0 in (10–15 cm)

Stride
Running: 0.7–3.0 ft (21–91 cm)

Size
Length with tail: 18–28 in (46–71 cm)

Weight
18–38 oz (0.5–1.1 kg)

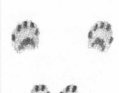

running

FOX SQUIRREL
Sciurus niger

Much like the Eastern Gray Squirrel, but larger and with a yellowish underside, this squirrel can be a common sight near the border with Minnesota, in deciduous forests where there are plenty of nut-bearing trees and in open areas near woodlots. Piles of nutshells at the bases of trees indicate the squirrel's favourite feeding sites. Active all year, this squirrel spends a lot of time foraging on the ground, often collecting the nuts that it buried singly during the previous fall.

The Fox Squirrel has a typical squirrel track when it runs or bounds, with the hind prints falling slightly ahead of the fore prints, each pair falling roughly side by side. A clear fore print shows four toes with evident claws, four fused palm pads and two heel pads. The hind print shows five toes, four palm pads and sometimes a heel.

Similar Species: Eastern Gray Squirrels (p. 90), in the south, leave smaller prints. Red Squirrels (p. 88) leave much smaller prints. A rabbit or hare (pp. 68–75) makes a longer print pattern; its fore prints are rarely side by side when it runs. Chipmunks (p. 94) and Northern Flying Squirrels (p. 86) leave smaller prints, with smaller straddles.

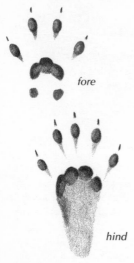

fore

hind

Fore Print
Length: 0.8–1.0 in (2.0–2.5 cm)
Width: 0.4–0.8 in (1.0–2.0 cm)

Hind Print
Length: 0.7–1.3 in (1.8–3.3 cm)
Width: 0.5–0.9 in (1.3–2.3 cm)

Straddle
2.0–3.1 in (5.1–7.9 cm)

Stride
Running: 7.0–15 in (18–38 cm)

Size
Length with tail: 7.0–10 in (18–25 cm)

Weight
2.5–5.0 oz (71–140 g)

running

EASTERN CHIPMUNK
Tamias striatus

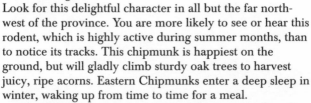

This large chipmunk is found in a variety of habitats, from the dense forest floor to open areas near buildings. Look for this delightful character in all but the far north-west of the province. You are more likely to see or hear this rodent, which is highly active during summer months, than to notice its tracks. This chipmunk is happiest on the ground, but will gladly climb sturdy oak trees to harvest juicy, ripe acorns. Eastern Chipmunks enter a deep sleep in winter, waking up from time to time for a meal.

Chipmunks are so light that their prints rarely show fine details. Each forefoot has four toes, while each hind foot has five. These chipmunks run on their toes, so their forefoot heel pads often don't register; the hind feet have no heel pads. The Eastern Chipmunk's erratic tracks are made up of hind prints registering ahead of fore prints, like those of many of their cousins. A chipmunk's trail often leads to extensive burrows.

Similar Species: The Least Chipmunk (*Tamias minimus*), found throughout the province except in the south, has smaller prints. The Red Squirrel (p. 88) has larger prints. Mice (pp. 98–101) prints are smaller. Mid-winter tracks are more likely to belong to squirrels than to chipmunks.

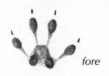

fore

hind

Fore Print
Length: 0.5 in (1.3 cm)
Width: 0.5 in (1.3 cm)

Hind Print
Length: 0.6 in (1.5 cm)
Width: 0.5–0.8 in (1.3–2.0 cm)

Straddle
1.3–2.0 in (3.3–5.1 cm)

Stride
Walking: 0.8 in (2.0 cm)
Running/Hopping:
    2.0–6.0 in (5.1–15 cm)

Size
Length with tail:
    5.5–8.0 in (14–20 cm)

Weight
0.5–2.5 oz (14–70 g)

walking

*running
(in snow)*

MEADOW VOLE
(Field Mouse)

Microtus pennsylvanicus

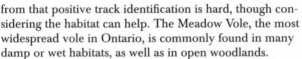

There are so many vole species to choose from that positive track identification is hard, though considering the habitat can help. The Meadow Vole, the most widespread vole in Ontario, is commonly found in many damp or wet habitats, as well as in open woodlands.

Vole tracks show four toes on the fore print and five on the hind, although it is hard to tell, because the prints are seldom clear. Voles have a paired alternating walking track, the hind print occasionally registering on the fore, but usually opt for faster leaping–the prints appear in hopping pairs, hind registering on fore. Voles stay under the snow in winter, so look for distinctive piles of cut grass from their ground nests and for tiny teeth marks in the bark at shrub bases. In summer, voles use the same paths so often that they appear as little runways through the grass.

Similar Species: The widespread Southern Red-backed Vole (*Clethrionomys gapperi*) is smaller. Rock Voles (*Microtus chrotorrhinus*) live in central and eastern regions, usually in rocky areas, while Heather Voles (*Phenacomys intermedius*) are further north. Southern Bog Lemmings (*Synaptomys cooperi*) leave similar tracks in grassy meadows, except in the far north. The Deer Mouse (p. 98) has a shorter stride.

running group

Fore Print
Length: 0.3 in (0.8 cm)
Width: 0.3 in (0.8 cm)
Hind Print
Length: 0.6 in (1.5 cm)
Width: 0.4 in (1.0 cm)
Straddle
1.4–1.8 in (3.6–4.6 cm)
Stride
Running: 2.0–5.0 in
    (5.1–13 cm)
Size
Length with tail:
    6.0–9.0 in (15–23 cm)
Weight
0.5–1.3 oz (14–37 g)

running

*running
(in snow)*

DEER MOUSE
Peromyscus maniculatus

The Deer Mouse,
one of the most
abundant mammals in
the region, is seldom seen
because it is nocturnal. This
highly adaptable rodent can be
found in most habitats, except for
dense woodlands and wet areas. It
remains active during the winter,
when it may shelter in buildings.

Deer Mouse fore prints each show four toes, three
palm pads and two heel pads. The hind prints show five
toes and three palm pads; the hind heel pads rarely
register. It takes perfect, soft mud to get clear prints from
such a tiny mammal. Running tracks–most noticeable in
snow–show the hind prints in front of the close-set fore
prints. The tracks may lead you up the occasional tree or
down into a burrow. In soft snow, the prints may merge
and appear as larger pairs of prints, with tail drag evident.

Similar Species: Many mice have near-identical tracks,
for example, the White-footed Mouse (*Peromyscus leucopus*)
of the south, the Meadow Jumping Mouse (p. 100), and the
House Mouse (*Mus musculus*), which is more associated with
humans. A vole's (p. 96) merged two-print patterns have
longer strides. Chipmunk (p. 94) tracks have a wider straddle. Shrew (p. 102) tracks have a narrower straddle.

running group

Fore Print
Length: 0.3–0.5 in (0.8–1.3 cm)
Width: 0.3–0.5 in (0.8–1.3 cm)

Hind Print
Length: 0.5–1.3 in (1.3–3.3 cm)
Width: 0.5 in (1.3 cm)

Straddle
1.8–1.9 in (4.6–4.8 cm)

Stride
Hopping: 2.0–7.0 in (5.1–18 cm)
In alarm: 3.0–4.0 ft (91–120 cm)

Size
Length with tail:
    7.0–9.0 in (18–23 cm)

Weight
0.6–1.3 oz (17–37 g)

running

MEADOW JUMPING MOUSE
Zapus hudsonius

Congratulations if you find and successfully identify the tracks of a jumping mouse! These rodents are hard to find, although they live throughout Ontario. The Meadow Jumping Mouse's preference for grassy meadows and dense undergrowth and its long, deep hibernation period (about six months!) makes locating tracks very difficult.

Jumping mouse tracks are distinctive if you do find them. The two smaller fore prints register between the long hind feet. The long heels do not always show. When they jump, these mice make short leaps. The tail may leave a dragline in soft mud or unseasonable snow. An abundant sign of this rodent is the clusters of cut grass stems, about 5 inches (13 cm) long, lying in the meadows.

Similar Species: The Woodland Jumping Mouse (*Napaeozapus insignis*), with similar tracks, is found in wet, wooded areas in most of Ontario except for the far north. Deer Mouse (p. 98) tracks may have the same straddle.

running group

Fore Print
Length and Width: 0.2 in (0.5 cm)

Hind Print
Length: 0.6 in (1.5 cm)
Width: 0.3 in (0.8 cm)

Straddle
0.8–1.3 in (2.0–3.3 cm)

Stride
Running: 1.2–2.3 in (3.0–5.8 cm)

Size
Length with tail:
    4.3–5.5 in (11–14 cm)

Weight
0.5–1.1 oz (14–31 g)

running

NORTHERN SHORT-TAILED SHREW

Blarina brevicauda

While there are many tiny, frenetic shrews in the province, if you find shrew tracks, the most likely candidate is the largest of the North American shrews, the Northern Short-tailed Shrew. Though widespread, it is absent from the far north. This shrew is just as happy in woodlands as it is around open marshes.

Its rapid activity makes it difficult to observe closely. In its energetic and unending quest for food, a shrew usually leaves a running pattern of four prints, but may slow to an alternating walking pattern. The individual prints in a group are often indistinct, but in mud or shallow, wet snow, you can even count the five toes on each print. In snow, a shrew's tail often leaves a dragline. A shrew's trail may disappear down a burrow. If the shrew tunnels under the snow, it may leave a ridge of snow on the surface.

Similar Species: Smaller shrew species have similar tracks. The secretive Masked Shrew (*Sorex cinereus*) shares similar habitats. The Arctic Shrew (*Sorex arcticus*) and the Water Shrew (*Sorex palustris*), both found in the north and the west, prefer moist habitats. The Smoky Shrew (*Sorex fumeus*) is found in southern and central regions. The Pygmy Shrew (*Sorex hoyi*) has considerably smaller prints. Mice (pp. 98–101) fore prints show only four toes.

A molehill of the Star-nosed Mole

Some molehills and ridges of the Eastern Mole

STAR-NOSED MOLE

Condylura cristata

This peculiar character is found throughout most of Ontario, except for the extreme northwest. The Star-nosed Mole is easily identified by the strange tentacle-like protrusions on its nose, thought to help it find food in its dark and often subterranean world. It spends more time out of its burrow than do other moles.

Typical signs of this mole include the big piles of mud pushed out of its burrows. Because of its preference for swimming and wet areas, look for this mole's hills along the banks of streams and rivers, and in raised areas around marshes and wet fields. Clear prints from its long-clawed feet will rarely be evident. The Star-nosed Mole remains under the snow during the winter and swims under the ice, so winter tracks are seldom seen.

Similar Species: The Eastern Mole (*Scalopus aquaticus*) occurs only near Lake Erie and leaves the familiar distinctive molehills often connected by raised ridges near the surface (as illustrated). The Hairy-tailed Mole (*Parascalops breweri*) can be found in central and southern Ontario. Both of these moles can be found in a wide diversity of habitats.

BIRDS, AMPHIBIANS AND REPTILES

A guide to the animal tracks of Ontario is not complete without some consideration of birds, amphibians and reptiles. Only a few have been chosen as examples.

The differences among bird species are not necessarily reflected in their tracks, so several bird species have been chosen to represent the main types common to this region.

Bird tracks can often be found in abundance in snow and are clearest in shallow, wet snow. The shores of streams and lakes are very reliable locations to find bird tracks—the mud there can hold a clear print for a long time. The sheer number of tracks made by shorebirds and waterfowl can be astonishing. While some bird species prefer to perch in trees or soar across the sky, it can be entertaining to follow the tracks of those that do spend a lot of time on the ground. They can spin around in circles and lead you in all directions. The track may suddenly end as the bird takes flight, or it might terminate in a pile of feathers, the bird having fallen victim to a hungry predator.

Many amphibians and turtles depend on moist environments, so look in the soft mud along the shores of lakes and ponds for their distinctive tracks. While you may be able to distinguish frog tracks from toad tracks, since they generally move differently, it can be very difficult to identify the species. In drier environments, reptiles outnumber amphibians, but dry terrain does not show prints well, except in sand. Snakes can leave distinctive tracks as they wind their way through mud or sand; foot-less, they make body prints.

Print
Length: to 1.5 in (3.8 cm)
Straddle
1.0–1.5 in (2.5–3.8 cm)
Stride
Hopping: 1.5–5 in (3.8–13 cm)
Size
5.5–6.5 in (14–17 cm)

DARK-EYED JUNCO
Junco hyemalis

This common small bird typifies the many small hopping birds found in this region. Each foot has three forward-pointing toes and one longer toe at the rear. The best prints are left in snow, although in deep snow the toe detail is lost, and the feet may show some dragging between the hops.

A good place to study these types of prints is near a birdfeeder. Watch the birds scurry around as they pick up fallen seeds, then have a look at the prints that they have left behind.

Similar Species: The exact dimensions of the toes may indicate what kind of bird you are tracking–larger birds will have larger footprints. Not all birds are present year-round, so keep in mind the season when tracking.

Print
Length: 2.5–3.0 in (6.4–7.6 cm)
Straddle
1.5–3.0 in (3.8–7.6 cm)
Stride
Walking: 4.0 in (10 cm)

AMERICAN CROW
Corvus brachyrhynchos

The black silhouette of the American Crow can be a common sight in all regions of Ontario, though it visits the north during the summer only. A crow will frequently come down to the ground and contentedly strut around with a confidence that hints at its intelligence. Its loud *caw* can be heard from quite a distance. Crows can be especially noisy when they are mobbing an owl or a hawk.

The American Crow typically leaves an alternating walking track. Each print shows three sturdy toes pointing forward and one toe pointing backward. When a Crow needs greater speed, perhaps in preparing for take-off, it will bound along, leaving irregular pairs of diagonally placed prints with a longer stride between each pair.

Similar Species: Jays make similar tracks. The Common Raven (*Corvus corax*) is much larger, with prints up to 4 inches (10 cm) long, with a stride of 6 inches (15 cm).

Print
Length: 2.0–3.0 in (5.1–7.6 cm)
Straddle
2.0–3.0 in (5.1–7.6 cm)
Stride
Walking: 3.0–6.0 in (7.6–15 cm)
Size
15–19 in (34–48 cm)

RUFFED GROUSE
Bonasa umbellus

   This ground-dweller prefers the quiet seclusion of coniferous forests in winter, so that will be the best place to find its tracks. If you follow them quietly you may be startled when the grouse bursts from cover underneath your feet. Its excellent camouflage usually affords it good protection. This widespread grouse is found throughout Ontario, except in the far north.

   The three thick front toes leave very clear impressions, but the short rear toe, which is angled off to one side, will not always show up so well. The neat, straight track of the Ruffed Grouse appears to reflect this bird's cautious approach to life on the forest floor.

Similar Species: The Spruce Grouse (*Dendragapus canadensis*), which has similar tracks, is absent from the far south but can be common in the thick undergrowth of the northern coniferous forests that it favours.

GREAT HORNED OWL
Bubo virginianus

This wide-ranging owl is often seen resting quietly in trees during the day, as it prefers to hunt at night. This owl is an accomplished hunter in snow, and the 'strike' that it leaves can be quite a sight if it registers well. The owl strikes through the snow with its talons, leaving an untidy hole, which is occasionally surrounded by imprints of wing and tail feathers. These feather imprints are made as the owl struggles to take off with possibly heavy prey. It is not the most graceful of walkers, preferring to fly away from the scene.

You might stumble across this owl's strike and guess that its target could have been a vole scurrying around underneath the snow. If you are a really lucky tracker, you will have been following the surface track of an animal to find that it abruptly ends with this strike mark, where the animal has been seized by an owl.

Similar Species: The Common Raven (*Corvus corax*) also leaves strike marks, usually with much sharper feather imprints.

Print
Length: 4.0–5.0 in (10–13 cm)
Straddle
5.0–7.0 in (13–18 cm)
Stride
Walking: 5.0–7.0 in (13–18 cm)
Size
2.7–4.0 ft (81–120 cm)

CANADA GOOSE
Branta canadensis

This common goose is a familiar sight in open areas near lakes and ponds. Its huge, webbed feet leave prints that can often be seen in abundance along the muddy shores of just about any waterbody, including those in urban parks, where the Canada Goose's green-and-white droppings can accumulate in prolific amounts.

The webbed feet of the Canada Goose each have three long toes that face forward. These toes register well, but the webbing between them does not always show on the print. The feet point inward, which give the bird a pigeon-toed appearance and may account for its waddling gait.

Similar Species: Many waterfowl, including Mallards (*Anas platyrhynchos*) and other ducks, leave similar prints. Exceptionally large ones are likely a swan's (*Cygnus* spp.).

Print
Length: 0.8–1.3 in (2.0–3.3 cm)
Stride
Erratic
Size
7.0–8.0 in (18–20 cm)

SPOTTED SANDPIPER
Actitis macularia

The bobbing tail of the Spotted Sandpiper is a common sight on the shores of lakes, rivers and streams, but you will usually find only one of these territorial birds in any given location. Because of its excellent camouflage, likely the first you will see of this bird will be when it flies away, its fluttering wings close to the surface of the water.

As it teeters up and down on the shore, it leave trails of three-toed prints. Its fourth toe is very small and faces off to one side at an angle. Sandpiper tracks can have an erratic stride.

Similar Species: All sandpipers and plovers, including the common Killdeer (*Charadrius vociferus*), have similar tracks, although there is much diversity in print size.

Print
Length: 1.5 in (3.8 cm)
Stride
Up to 1.3 in (3.3 cm)
Size
11–12 in (28–30 cm)

COMMON SNIPE
Gallinago gallinago

This short-legged character is a resident of marshes and bogs, where its neat prints can often be seen in mud. Snipes are quite secretive when on the ground, and so you may be surprised as one suddenly flushes out from beneath your feet. Watch for the occasional snipe perched on a snag or a fence post, and listen for the eerie whistle as a snipe dives from the sky.

The Common Snipe's neat prints show four toes, including a small rear toe that points inward. The bird's short legs and stocky body give it a very short stride.

Similar Species: Many shorebirds, including the sandpipers, leave tracks like those of the Common Snipe.

Print
Length: to 6.5 in (17 cm)
Stride
9.0 in (23 cm)
Size
4.2–4.5 ft (1.3–1.4 m)

GREAT BLUE HERON
Ardea herodias

The refined and graceful image of this large heron symbolizes the precious wetlands in which it patiently hunts for food. Still and statuesque as it waits for a meal to swim by, the Great Blue Heron will walk from time to time, perhaps to find a better hunting location. Look for its large, slender tracks along the banks or mudflats of waterbodies.

Not surprisingly, a bird that lives and hunts with such precision walks in a similar fashion, leaving straight tracks that fall in a nearly straight line. Look for the slender rear toe in the print.

Similar Species: Cranes have similar habitats and similar prints, but their smaller hind toes do not register.

Straddle
to 2.5 in (6.4 cm)

TOADS AND FROGS

The best place to look for toad and frog tracks is undoubtedly along the muddy fringes of waterbodies, but toad tracks can occasionally be found in drier areas, for example as unclear trails in dusty patches of soil. In general, toads walk and frogs hop, but toads are pretty capable hoppers, too, especially when being hassled by overly enthusiastic naturalists.

TOADS

There are fewer toads than frogs in Ontario. The toad most likely to be found, and the most wide-spread, is the American Toad (*Bufo americanus*), which lives in many different moist habitats. Woodhouse's Toad (*Bufo woodhousei*) is common in the United States, but also extends just into southern Ontario; it enjoys temporary pools and ditches. Ontario's toads can be up to 4.5 inches (11 cm) in length.

American Toad

Toads leave rather abstract prints as they walk–the heels of the hind feet do not register. In less firm surfaces, you can often see the draglines left by the toes.

125

Straddle
to 3.0 in (7.6 cm)

FROGS

Frog tracks vary greatly in size, depending on species and age. The smallest frogs include the treefrogs, such as the Spring Peeper (*Hyla crucifer*). Their

Northern Leopard Frog

small size–a length of just 1.5 in (3.8 cm)–and their preference for thick undergrowth and shrubs by water means that their tracks are a rare sight. The larger Gray Treefrogs (*Hyla chrysoscelis* and *Hyla versicolor*) spend most of their time in trees, coming down to breed and sing in the night.

Even larger frogs include the widespread Woodfrog (*Rana sylvatica*), which is about 3.5 inches long (8.9 cm) and can often be found in dry woodland areas, seemingly far from water. The Green Frog (*Rana clamitans*) and Pickerel Frog (*Rana palustris*) are also widespread.

A frog's hopping action results in the two small fore prints registering in front of the long-toed hind prints. Larger tracks might come from the beautiful and widespread Northern Leopard Frog (*Rana pipiens*), which can grow up to 5 inches (13 cm) long.

An unusually large track is surely from the robust Bullfrog (*Rana catesbeiana*). At up to 8 inches (20 cm) in length, it is the largest frog in North America.

NEWTS AND SALAMANDERS

Red-Backed Salamander

There are a wealth of newts and salamanders in the region—more in the south than in the far north. These long, slender lizard-like amphibians are at home in moist or wet areas; look under logs near waterbodies and in moist woodlands. Look for tracks in the soft mud of a woodland path.

Among the more abundant and widespread species is the Eastern Newt (*Notophthalmus viridescens*), which can be 5.5 inches (14 cm) long. After a fresh rain, Eastern Newts emerging from ponds may leave small trails in the mud.

There are more species of salamanders than newts. In the mixed or coniferous forests of central and southern Ontario, tracks might be from the Red-backed Salamander (*Plethodon cinereus*), which can be 5 inches (13 cm) in length. Ontario's largest salamander is the stunning Spotted Salamander (*Ambystoma maculatum*), which has vibrant yellow spots on its dark body and measures to 10 inches (25 cm) in length. Its tracks, with a straddle of about 3 inches (7.6 cm), are rare, as it prefers to stay underground.

As a general rule for newts and salamanders, the fore print shows four toes, while the larger hind print shows five. However, print detail is often obscured by a dragging belly or by the swinging of a thick tail across the track.

SKINKS

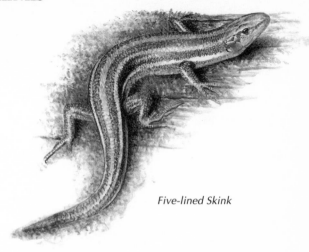

Five-lined Skink

Skinks leave tracks similar to those of salamanders, but their toes are longer and more slender. The most likely candidate is the Five-lined Skink (*Eumeces fasciatus*), which can grow up to 8 inches (20 cm) in length. It likes the moist woodlands of southern Ontario and favours the tops of fallen tree trunks. For the best tracks—created when a skink has moved between trunks or to a resting place—look in sandy soils around fallen trees.

Skinks can move very quickly when the need arises, their feet barely touching the ground as they dart for cover. Consequently, their tracks can be hard to make out clearly.

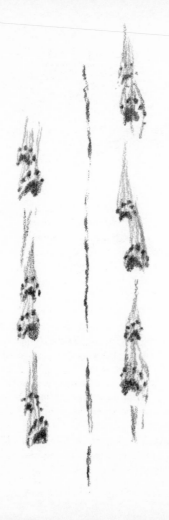

TURTLES

Wood Turtle

Turtles, those ancient
denizens of the water world, will
happily slip into the murky depths of the
water to avoid detection. They do, however, come out from
time to time to feed or bask in the sunshine. Look for their
distinctive tracks in the southern regions of the province,
along the banks of ponds and rivers and in other moist areas. Note that some turtles, such as the huge Snapping Turtle (*Chelydra serpentina*), prefer to stay in the water and
rarely come out.

The Map Turtle (*Graptemys geographica*) grows to about
10 inches (25 cm) in length. A slightly smaller turtle, the
Wood Turtle (*Clemmys insculpta*), is often associated with
woodland streams, but only in the south, in the Windsor
area and near the Quebec border.

With such short legs, a turtle leaves a track that is wide
relative to the length of its stride—its straddle is about half
the length of its body. Although longer-legged turtles can
raise their shells off the ground, those with short legs may
let their shells drag, as shown in their tracks. The tail may
leave a straight dragline in mud. On firmer surfaces, the
claws may register.

SNAKES

*Common Garter
Snake*

There are many snakes to be found through most of the province, with much greater diversity in the warmer south. As all snakes are long and slender, their tracks appear so similar that identification among the many species is next to impossible. In fact, because a snake lacks feet and leaves a track that is just a gentle meander, it is very challenging even to establish in which direction the snake is travelling.

The most widespread and frequently encountered snake—and the only one to be found in more northerly regions—is the Common Garter Snake (*Thamnophis sirtalis*). It can be found throughout the province, except in the far north, often close to wet or moist areas. This harmless snake can reach 4.3 feet (1.3 m) in length.

Here are three other widespread species: The small Red-bellied Snake (*Storeria occipitomaculata*), which grows to only 16 inches (41 cm) in length, prefers woodland. The Smooth Green Snake (*Opheodrys vernalis*) is only slightly larger. Watch out for the larger snap-happy Northern Water Snake (*Nerodia sipedon*). In sandy, open areas in the south, a track might well be from the Eastern Hognose Snake (*Heterodon platyrhinos*).

TRACK PATTERNS & PRINTS

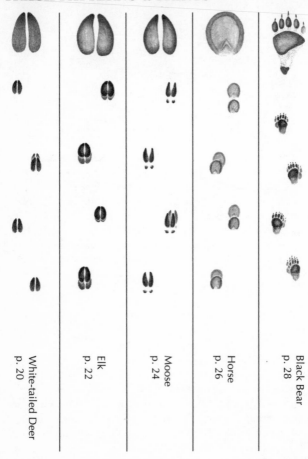

White-tailed Deer
p. 20

Elk
p. 22

Moose
p. 24

Horse
p. 26

Black Bear
p. 28

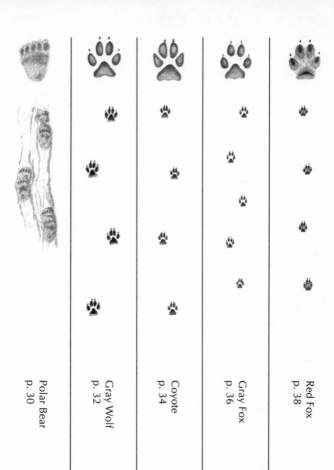

Polar Bear
p. 30

Gray Wolf
p. 32

Coyote
p. 34

Gray Fox
p. 36

Red Fox
p. 38

TRACK PATTERNS & PRINTS

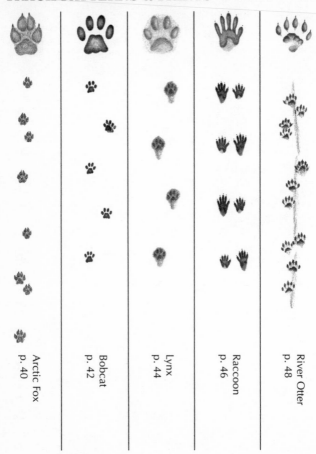

Arctic Fox
p. 40

Bobcat
p. 42

Lynx
p. 44

Raccoon
p. 46

River Otter
p. 48

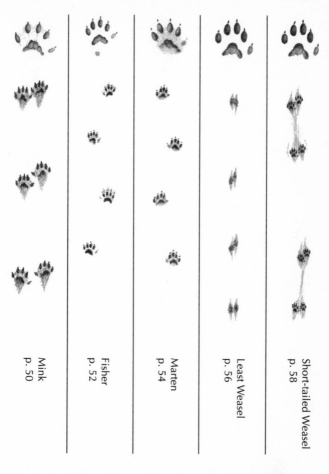

Mink
p. 50

Fisher
p. 52

Marten
p. 54

Least Weasel
p. 56

Short-tailed Weasel
p. 58

TRACK PATTERNS & PRINTS

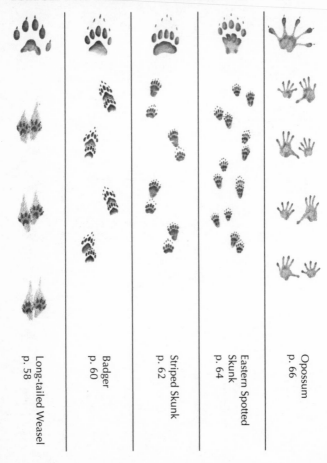

Long-tailed Weasel
p. 58

Badger
p. 60

Striped Skunk
p. 62

Eastern Spotted
Skunk
p. 64

Opossum
p. 66

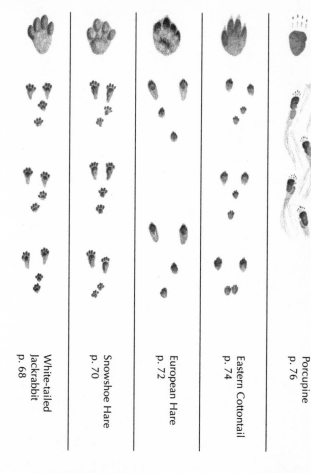

White-tailed
Jackrabbit
p. 68

Snowshoe Hare
p. 70

European Hare
p. 72

Eastern Cottontail
p. 74

Porcupine
p. 76

TRACK PATTERNS & PRINTS

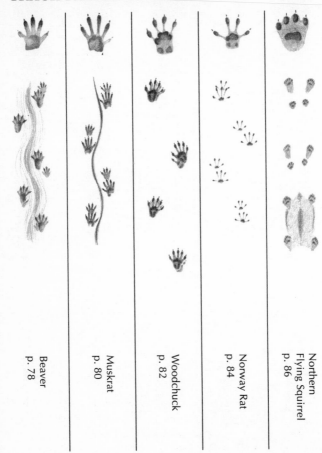

Beaver
p. 78

Muskrat
p. 80

Woodchuck
p. 82

Norway Rat
p. 84

Northern
Flying Squirrel
p. 86

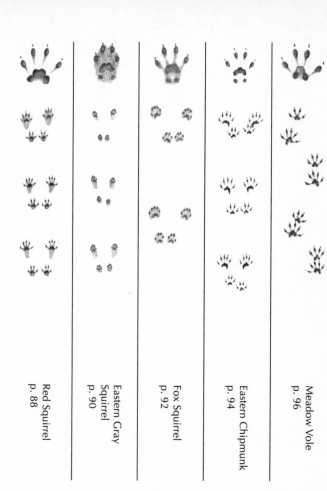

Red Squirrel
p. 88

Eastern Gray
Squirrel
p. 90

Fox Squirrel
p. 92

Eastern Chipmunk
p. 94

Meadow Vole
p. 96

TRACK PATTERNS & PRINTS

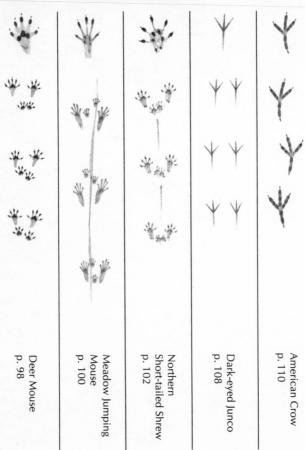

Deer Mouse
p. 98

Meadow Jumping
Mouse
p. 100

Northern
Short-tailed Shrew
p. 102

Dark-eyed Junco
p. 108

American Crow
p. 110

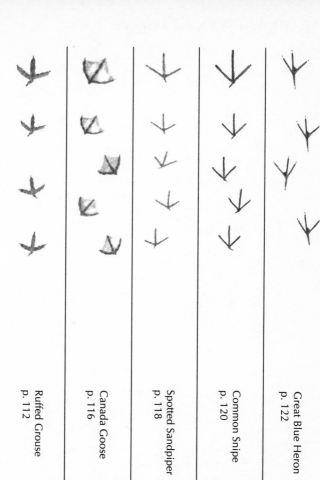

Ruffed Grouse
p. 112

Canada Goose
p. 116

Spotted Sandpiper
p. 118

Common Snipe
p. 120

Great Blue Heron
p. 122

Toads
p. 124

Frogs
p. 126

Newts and
Salamanders
p. 128

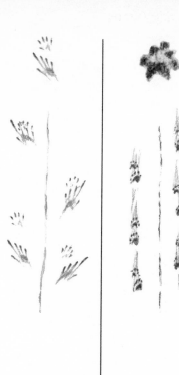

Skinks
p. 130

Turtles
p. 132

Snakes
p. 134

 147

HOOFED PRINTS

White-tailed Deer

Elk

Moose

Horse

FORE PRINTS

Least Weasel

Short-tailed Weasel

Long-tailed Weasel

Mink

Striped Skunk

Eastern
Spotted Skunk

Fisher

River Otter

Marten

Badger

```
inch      cm
0         0

1

2         5
```

FORE PRINTS

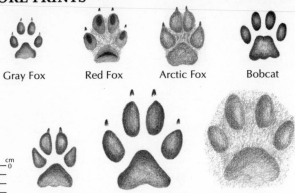

Gray Fox   Red Fox   Arctic Fox   Bobcat

inch cm
0 — 0
1
2 — 5

Coyote   Gray Wolf   Lynx

HIND PRINTS

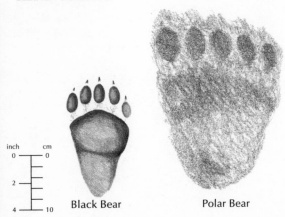

inch cm
0 — 0
2
4 — 10

Black Bear   Polar Bear

HIND PRINTS

Opossum

Raccoon

Muskrat

Woodchuck

White-tailed
Jackrabbit

Snowshoe
Hare

European
Hare

Eastern
Cottontail

Porcupine

Beaver

inch   cm
0   0

1

2   5

 151

HIND PRINTS

Deer Mouse

Northern
Short-tailed
Shrew

Meadow
Jumping Mouse

Meadow
Vole

Eastern
Chipmunk

Norway Rat

Red Squirrel

Fox Squirrel

Northern
Flying Squirrel

Eastern Gray
Squirrel

inch cm
0 — 0

1 —

2 — 5

BIBLIOGRAPHY

Behler, J.L. and F.W. King. 1979. *Field Guide to North American Reptiles and Amphibians.* National Audubon Society. New York: Alfred A. Knopf, Inc.

Burt, W.H. 1976. *A Field Guide to the Mammals.* Boston: Houghton Mifflin Company.

Farrand, J., Jr. 1995. *Familiar Animal Tracks of North America.* National Audubon Society Pocket Guide. New York: Alfred A. Knopf, Inc.

Forrest, L.R. 1988. *Field Guide to Tracking Animals in Snow.* Harrisburg: Stackpole Books.

Halfpenny, J. 1986. *A Field Guide to Mammal Tracking in North America.* Boulder: Johnson Publishing Company.

Headstrom, R. 1971. *Identifying Animal Tracks.* Toronto: General Publishing Company, Ltd.

Jones, J.K., Jr. and E.C. Birney. 1988. *Handbook of Mammals of the North-Central States.* Minneapolis: University of Minnesota Press.

Murie, O.J. 1974. *A Field Guide to Animal Tracks.* The Peterson Field Guide Series. Boston: Houghton Mifflin Company.

Rezendes, P. 1992. *Tracking and the Art of Seeing: How to Read Animal Tracks and Sign.* Vermont: Camden House Publishing, Inc.

Stall, C. 1989. *Animal Tracks of the Great Lakes.* Seattle: The Mountaineers.

Stokes, D. and L. Stokes. 1986. *A Guide to Animal Tracking and Behaviour.* Toronto: Little, Brown and Company.

Whitaker, J.O., Jr. 1996. *National Audubon Society Field Guide to North American Mammals.* New York: Alfred A. Knopf, Inc.

INDEX

Page numbers in **boldface** type refer to the primary
(illustrated) treatments of animal species and their tracks.

NOTES

Look for these Ontario titles from Lone Pine Publishing

To order: 1-800-661-9017 phone • 1-800-424-7173 fax

Ontario Birds
Vivid de... ...ommon
and inte... ...olor
illustrati... ...out.
5.5" x 8.5" • 160 pgs • softcover
$17.95 CDN • $14.95 US
1-55105-069-2

City Bird Guide Series
Species descriptions, full-color
illustrations, directions to the best
birdwatching locations and more.
5.5" x 8.5" • 144 pgs • softcover
$11.95 CDN • $9.95 US
Birds of Ottawa 1-919433-64-2
Birds of Toronto 1-919433-63-4

Forest Plants of Northeastern Ontario
More than 300 species of trees, shrubs, wildflowers,
grasses, ferns, mosses and lichens. More than 300
color photos, 330 line drawings.
4.25" x 8" • 352 pgs • softcover • $24.95 CDN • $19.95 US
1-55105-064-1

Forest Plants of Central Ontario
Over 390 species of trees, shrubs,
wildflowers, grasses, lichens and ferns.
440 color photos, 407 line drawings.
4.25" x 8" • 448 pgs • softcover
$24.95 CDN • $19.95 US
1-55105-061-7

Wetland Plants of Ontario
More than 450 species, with 300
color photos and 300 line
drawings throughout.
5.5" x 8.5" • 256 pgs • softcover
$24.95 CDN • $19.95 US
1-55105-059-5